Hibernation

Paul Bennett

Thomson Learning
New York

Nature's Secrets

Cover: A common dormouse hibernating.

Title page: A brown bear has awakened from its winter sleep to feed.

Contents page: A red squirrel finds food among the autumn leaves.

First published in the
United States in 1995 by
Thomson Learning
115 Fifth Avenue
New York, NY 10003

Published in Great Britain in 1994 by
Wayland (Publishers) Ltd.

Library of Congress Cataloging-in-Publication Data
Bennett, Paul, 1954–
  Hibernation / Paul Bennett.
      p.   cm.—(Nature's secrets)
  Includes bibliographical references (p.   ) and
index.
  ISBN 1-56847-208-0
  1. Hibernation—Juvenile literature.
[1. Hibernation.] I. Title. II. Series: Bennett,
Paul, 1954–   Nature's secrets.
QL755.B45   1994
591.54'3—dc20   94-27584

Printed in Italy

Picture acknowledgments
The publishers would like to thank the following for allowing their photographs to be reproduced in this book: Bruce Coleman Ltd. *cover* (George McCarthy), *title page* (Francisco Marquez), 4 (Jeff Foott), 6 (Wildtype Prods.), 7 (top), 12, 13 (top) , 21 (Andy Purcell), 10 (Erwin & Peggy Bauer), 11 (Jen & Des Bartlett), 13 (bottom/John Fennell), 16 (top/Robert P. Carr), 17 (Wayne Lankinen), 19 (top/Fred Bruemmer), 19 (bottom/Hans Reinhard), 20 (Mike Price), 23 (top/Jane Burton), 25 (top/P. Clement), 27 (bottom/Bob & Clara Calhoun), 29 (top/O. Langrand); Frank Lane Picture Agency *contents page* (R. Bender), 8 (R. Wilmshurst); Natural History Photographic Agency 9 (Stephen Dalton), 22 (Anthony Bannister), 24 (Otto Rogge © A.N.T.), 25 (bottom/George Gainsburgh), 26 (top/David Woodfall), 26 (bottom/G. I. Bernard), 27 (top/George Gainsburgh/G. I. Bernard); Oxford Scientific Films 5 (both/Norbert Rosing), 7 (bottom/Robert A. Tyrell) 14 (top/Richard Packwood), 14 (bottom/Robert A. Lubeck), 15 (Ken Highfill, Photo Researchers Inc.), 16 (bottom/Tom McHugh, Photo Researchers Inc.), 18 (Zig Leszczynski, Animals Animals), 23 (bottom/G. I. Bernard), 28 (both/Jim Frazier, Mantis Wildlife Films), 29 (bottom/Fran Allan, Animals Animals).

Contents

Surviving the cold

Animals have different ways of
surviving harsh winter weather.
Some are able to stay active in order
to search for food. The Arctic fox stays
awake and on the move through the
icy weather of the northern winter.
It has a thick coat of soft fur to help it
keep warm. ▽

Others escape the cold weather by
going into a kind of deep sleep called
hibernation. They awaken when the
warm weather returns and food is
plentiful again.

△ A female polar bear digs a den in the snow and rests there through the winter suckling her newborn cubs. In the spring, the cubs are strong enough to follow their mother when she leaves the den, looking for food.

5

△ In winter, the crayfish buries itself
in mud at the bottom of a pond and
becomes sleepy, or torpid. Although a
pond may have a top covering of ice,
the water at the bottom is not as
cold, so the animals living there will
not freeze.

△ Adders and other snakes
sleep through the cold season.
They have to find a place
where they will not freeze.

Most birds do not hibernate.
But some, such as dazzling
hummingbirds, become
torpid at night. As the
temperature drops, their
body functions slow down
to conserve energy. ▷

Mammals

When people sleep their heartbeat and breathing are slower than when they are awake and moving around. Hibernation is similar to sleep, but there are greater changes in the animal's body. Its winter sleep is so deep that it is almost impossible to wake. It may appear to be dead. In this state the animal uses up very little energy and is able to survive the winter without feeding. There are few true hibernators, and the dormouse is one of them.

◁ During the summer and autumn months a dormouse prepares for winter by feeding greedily on fruits and seeds. Soon it builds up a store of fat that provides the energy it needs for hibernation.

After leaving its summer nest the dormouse will make a cosy winter nest among tree roots or underground. Inside, among the leaves and grass, it will be snug and safe for its half year of sleep.

While hibernating, the dormouse may breathe only once every few minutes, its heartbeat becomes very weak, and its temperature drops until it is almost as cold as its surroundings. In the spring, it wakes up and returns to its normal activities.

There are other mammals that scientists once thought might be true hibernators. One of these is the North American opossum. Although it slows its activity in winter, it is not a true hibernator.
▽

The Australian koala also seems to spend part of the year hibernating, but it is not a true hibernator either. ▷

Only some small mammals are true hibernators. Larger mammals may sleep during the cold weather, but they can be awakened from their dozy state by a loud noise. These mammals may also come out in warmer periods to search for food.

Mammals are warm-blooded animals. This means that their body temperature does not depend on their surroundings, and under most conditions they are able to maintain their normal body temperature.

In autumn, the golden-mantled ground squirrel disappears into a burrow to hibernate. Many true hibernators have a store of food that they will eat when they wake up in spring. ▽

△ The European hedgehog hibernates in a bed of leaves or straw in a sheltered place. It has fed well and will sleep deeply until the weather becomes mild and the slugs and insects on which it feeds become abundant again.

In Australia, the spiny ▷ anteater hibernates for several months at a time. However, it does wake up occasionally to feed.

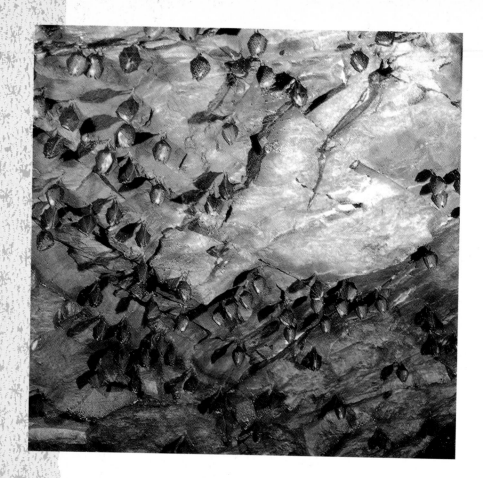

◁ For lesser horseshoe bats, a damp cave is an ideal place to hibernate. They cluster together to conserve heat and become covered by water droplets from condensation.

Squirrels are found in both Europe and the United States. This gray squirrel is snug and warm in its nest. It will remain here for days at a time and will only venture out once in a while to look for food. ▷

◁ An American red squirrel looks out of its snow-covered nest. During the autumn, it buries seeds in the ground that it will eat during the cold winter months.

◁ Skunks do not hibernate, but they may sleep for days during bitterly cold weather, becoming active again on warmer days.

Here is a sleeping chipmunk. These ground-living squirrels do not have the body reactions unique to true hibernators. ▽

Brown bears and black bears sleep in dens when the weather gets cold. Like polar bears, these bears produce their cubs during the winter and come out of their dens in the spring.

Snakes and lizards

Unlike mammals and birds, snakes and lizards cannot keep themselves active and warm in winter. They are usually about as warm as their surroundings, and the colder they get, the less active they become. Animals that make little heat in their bodies and tend to have the temperature of their surroundings are called "cold-blooded."

△ Torpid garter snakes cluster together in order to share the warmth of their bodies. They find a hole in the ground or a place among rocks where they can hibernate through the winter until the warm weather returns.

With the return of the warm spring weather, these garter snakes wriggle out of their rocky winter hiding place. ▷

◁ A green lizard warms itself in the sun. After waking from their winter sleep, snakes and lizards can be seen basking in the sun to warm themselves before they hunt for food.

Tortoises and terrapins

Tortoises and terrapins are cold-blooded creatures. In cooler parts of the world they must find a safe place to sleep during the winter.

△ The North American diamondback terrapin spends the winter buried in the mud of a pond or lake. Here it remains torpid, conserving energy.

△ Pet tortoises hibernate, too. When the
autumn comes this girl will put her tortoise in
a straw-filled box and leave it in a frost-free
place until the warm weather returns.

Frogs, toads, and newts

Frogs, toads, and newts belong to a
group of cold-blooded animals called
amphibians. To survive cold weather,
they creep into holes in the ground,
or under large stones or logs, and go
into a deep sleep.

△ As autumn approached, this bullfrog
ate plenty of food to build up a supply of
fat in its body. Now it can draw upon
this source of energy as it sleeps.

This toad is hibernating with some snails in a hole. Its heart is beating just enough to circulate the blood throughout its body. ▷

A great crested newt leaves the water to find a place to hibernate. ▽

Invertebrates

Many other types of animals cannot make heat to warm themselves when the weather turns cold. Some of them hibernate. In some insect species, most of the adults die in the autumn, leaving only a few to sleep through the winter and breed the following spring.

△ Snails prefer damp places in which to hibernate. They go into their shells and seal up the opening so that they do not lose water.

△ Slugs creep underground or into rock crevices. They can sleep without feeding for several months until better weather returns.

Wood lice shelter in a crevice where they are protected from the harsh weather. ▽

◁ A small tortoiseshell butterfly hibernates indoors. Some butterflies hide under logs, in hollow trees, or even in houses.

This queen wasp has found a quiet place in which to sleep. In the spring, she will lay her eggs and start a new nest. ▷

△ A bumblebee queen emerges from hibernation. Soon she will fly away to find a burrow in the ground in which to nest.

Ladybugs cluster on plants to survive the bitter weather. Some insects can make a chemical in their bodies that prevents them from freezing. ▷

When it gets hot

In warm countries, some animals escape the hottest and driest times of the year in a state called estivation, which is a similar sort of sleep to hibernation.

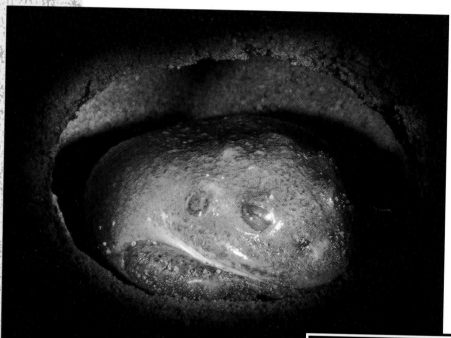

◁ The Australian burrowing frog makes an underground chamber to escape the heat. It sleeps in a waterproof skin, which prevents it from drying out.

The frog removes the skin and emerges from its burrow as soon as the rains come. ▷

△ The dwarf lemur becomes torpid so that it can survive the dry season.

Even some fish estivate. The lungfish swims happily in a pool, but as soon as the water dries up, it will burrow into the mud where it remains breathing air until the drought is over. This lungfish has been dug out of its muddy resting place. ▷

Glossary

Amphibians Cold-blooded animals that live on land but breed in water.

Cold-blooded Animals that are unable to make heat to warm their body. Their body temperature is similar to that of their surroundings.

Condensation Drops of water formed from the air.

Conserve To keep from being wasted.

Dormant Sleeping. Not active.

Drought A period of time when no rain falls.

Emerge To come out.

Hibernation The deep sleep of some small mammals, during which bodily functions slow down. The word is also used to describe the long sleep of any animal during the winter months.

Insect Small, six-legged creature with a body divided into three separate parts. Many insects have one or two pairs of wings.

Invertebrates Animals that have no backbone. Insects, spiders, clams, worms, and sponges are all examples of invertebrates.

Mammals Animals whose females give birth to live young that are fed with milk from the mothers' bodies.

Suckling A young animal that feeds on its mother's milk.

Torpid Extremely sleepy and barely able to move.

Warm-blooded Animals that are able to make heat from their bodies. They are able to maintain a constant body temperature whether their surroundings are warm or cold.

Books to read

Bailey, Donna. *Bears*. Animal World. Milwaukee: Raintree Steck-Vaughn, 1990.

Brimner, Larry Dane. *Animals that Hibernate*. First Books. New York: Franklin Watts, 1991.

Shebar, Sharon Siegmond and Shebar, Susan E. *Bats*. First Books. New York: Franklin Watts, 1990.

Tesar, Jenny. *Mammals*. Our Living World. Woodbridge, CT: Blackbirch Press, 1993.

Projects

Project: **Hibernating**

In the summer, gardens and parks are full of wildlife. During the cold winter months our yards look dead and deserted. However, some of the animals that visit your yard or park may stay to hibernate. If you or a neighbor have a pond, frogs, toads, or other amphibians may hibernate there. Some butterflies hibernate as adults, others as chrysalises.
You may find dormant butterflies in a garden shed or in your attic. Some snakes and toads hibernate in compost heaps. You may find a hibernating frog or toad in your yard. Keep a record of any hibernating animals that you find. If you have a pet tortoise or turtle, you may be able to observe it as it hibernates through the winter. Never touch or disturb a hibernating animal because it may awaken and die.

Project: **Staying Awake**

Many animals stay awake through the cold winter months. You will have a better chance of seeing these animals if you put food out for them. A selection of bird food on a bird table will attract a variety of birds. If you do not have a yard, you can put bird food on a window sill. Also put out water—birds need to drink and bathe all year long. Nuts may also attract squirrels, which stay active in the winter. Keep a note of the different types of birds and animals that visit your yard. Record the different types of food that they eat. When do they visit, and how long do they stay? Make a wildlife diary.

If it snows, look for animal tracks in your yard or local park. You may find footprints made by squirrels, rabbits, birds, or even a fox. Draw the shape of the print and try to identify the animal from a book. Write down all your observations and then add your drawings to your wildlife diary.

Index